# Miss Penny Always Says PROVE IT!

A Math-Infused Story about
Developing Number Knowledge and Exercising
the Mathematical Practice Standards.

## Lynda Brennan

ILLUSTRATED BY
### Richard H. Walsh

**Math MileMarkers®**
Math MileMarkers® Educational Book Series
Math4Minors.com • MathMileMarkers.com

**ACKNOWLEDGMENTS**

This document contains common core content standards and mathematical practice standards from the Common Core State Standards for Mathematics.

Additional content quoted from *Teaching Math to Young Children: A Practice Guide* (NCEE 2014-4005) by the National Center for Education Evaluation and Regional Assistance (NCEE), Institute of Education Sciences, U.S. Department of Education.

Cover design and interior illustrations by Richard H. Walsh
Book design by Lynn Else

ISBN: 979-8-9870497-2-3

Library of Congress Cataloging-in-Publication Data is available upon request.

Printed and bound in the
United States of America

*Dedication*

To my family, thank you…dreams do come true!


*Special Acknowledgment*

Special thanks to all my collegues, students and to the very talented young artists who contributed their rendition of the number 16 for display in this book: Matthew, Christopher, Rosa, Emily, Molly, Olivia E., Jessica, Olivia G., Kailyn, Olivia M., Gianna, John, Christopher, Ivannia, Jimmy, John, and Maggie.

Miss Penny

K-1

4

Miss Penny smiled as her class barreled
into her classroom, ready for another day.
The students loved Miss Penny's room,
and they loved her.

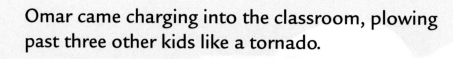

Omar came charging into the classroom, plowing past three other kids like a tornado.

"Oh, Omar," squeaked Miss Penny.

Just then, Bobby stumbled into the room. "What's up, Bobby, are you OK?" asked Miss Penny.

Bobby pointed to his back pocket. There was a giant bulge popping out, clearly not something he could sit on for an entire school day.

"What's in your pocket?" asked Miss Penny. Bobby excitedly reached back to reveal his prized marble collection.

Miss Penny held out both of her hands as Bobby grabbed a handful of marbles from his back pocket and placed them carefully within her grasp.

"They are beautiful!" she exclaimed.

Miss Penny then walked cautiously to the carpet, her hands held out in front of her so as not to drop a single one. The entire class followed her.

"This is a perfect way to start our day," said Miss Penny. "Can anyone guess how many marbles Bobby put in my hands?"

Rosa whispered, "Two?"

Maggie, with her arms folded confidently, said, "Twenty."

Sammy screamed, "Eleven."

Jimmy giggled and said, "Ninety-nine," while John chimed in with, "No, one hundred marbles!"

"Wow," said Miss Penny, "one hundred is a really big number." Everyone in the room offered his or her best guess. Their answers varied greatly.

"Well, there's only one way to know for sure.
Let's count them together," said Miss Penny.
The children then gathered in a circle and
counted along as Bobby took each marble,
one by one from Miss Penny's hands,
and placed them in a pile on
the floor.

"One, two, three, four, five, six, seven, eight, nine, ten, eleven, twelve, thirteen, fourteen, fifteen, sixteen."

Miss Penny shook out her hands, relieved to have them empty again, and asked Bobby, "How many marbles did we count?"

"Sixteen marbles," he responded proudly.

"That's a very nice collection, Bobby."

"We certainly can't leave them on the floor," said Miss Penny. "Omar, can you go get that large blue basket on the back table and help me put Bobby's marbles inside?"

Omar brought the basket back to the rug and placed all the marbles inside, handfuls at a time.

The marbles that once fit nicely in Miss Penny's hands were now spread out across the bottom of the blue basket, making the collection seem larger than ever.

"Well, that's interesting," said Miss Penny as she watched the marbles filter across the bottom of the basket.

"How many marbles do you think we have now?" The children all offered up their best estimate. Their guesses filled the air: "thirty-five, four, sixty-six, twenty-five, forty-five, seventy."

Once again their answers varied greatly. Now that the marbles were spread across the bottom of the large basket, surely there must be more than sixteen, was the comment from several students.

"Well, there's only one way to know for sure; let's count them and prove it!" declared Miss Penny.

Maggie came to the front to count. She pulled the marbles from the basket one by one and placed them in the center of the crowd.

Maggie looked back at the class and said, "I'm going to count them in groups of ten." Putting the marbles in a straight line, Maggie began to count.

"One, two, three, four, five, six, seven, eight, nine, ten. That's ten," she said. Starting a new row, she went on to say, "One, two, three, four, five, six. That's six in this row. Ten plus six more equals sixteen."

"I like the way you grouped those marbles, Maggie," said Miss Penny. "Can anyone show me another way to count and prove that we have sixteen marbles? That's right, prove it!"

Regina stepped forward and said
"I can count the marbles in groups of two.
Two, four, six, eight, ten, twelve, fourteen, sixteen.
It's definitely sixteen."

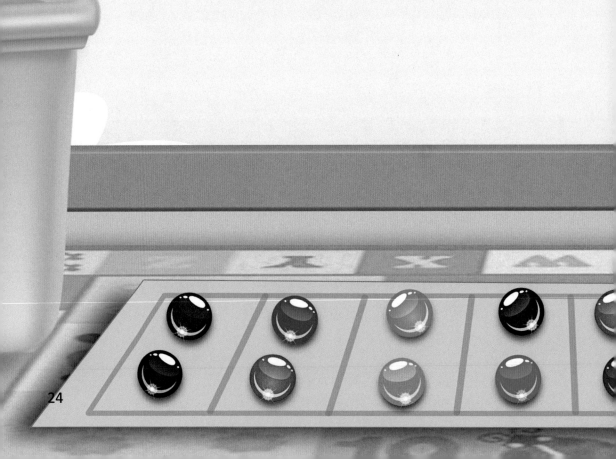

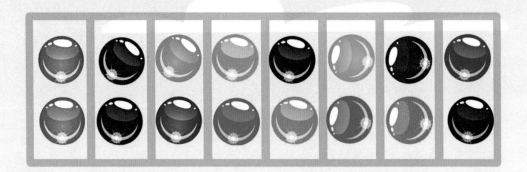

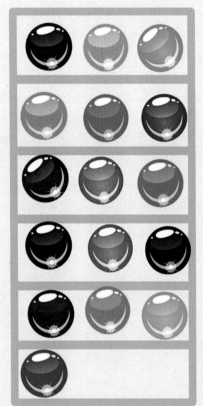

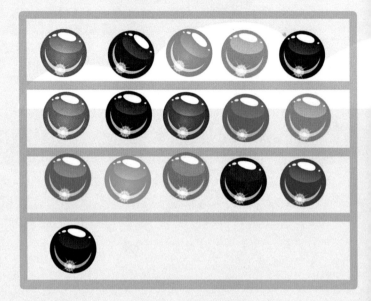

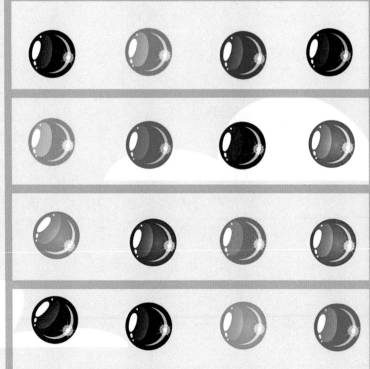

"Very good," said Miss Penny. "So what you learned is that sixteen marbles is sixteen marbles regardless of whether the marbles are in a small space or a big space.

Sixteen marbles will always have the same value: sixteen. It doesn't matter how we count them, in piles or in groups of ten and some extra, sixteen will always equal sixteen."

Miss Penny took the marbles from the floor. This time she put them in a small lunch bag and handed them back to Bobby.

"Thank you so much for bringing your marble collection into school today, Bobby, that was so much fun. These marbles have been in your pocket, in my hands, in the big basket, and now they are in this lunch bag. Tell me, Bobby, how many marbles do you think are in the bag now?"

With a giant smile and a whole lot of confidence, Bobby answered, "Sixteen!"

Sixteen became the number of the day. The students joyfully returned to their seats with sixteen on their minds.

It was their turn to prove it by showing the value of the number sixteen in their own special way.

31

The following conversation prompts and teacher notations are offered to help launch rich math conversations. Appropriate for both full-class and small-group discussions, these prompts provide an overview of the important content, mathematical practice standards, and the vocabulary presented in the story.

**BIG IDEA**   The Learning Environment

How do you think the children in this story feel about learning in Miss Penny's classroom?

Describe your perfect learning environment. Some things to consider: quiet or loud? Inquiry-based lessons or scripted? Allow children to use their own words to describe what a classroom utopia might look like.

How can we as a class work to create a positive learning environment for everyone in our class?

**BIG IDEA**   Open Counting Opportunities/Making Personal Connections

Bobby enters the room with a collection of marbles that he owns, not a prescribed number of items to count. How might this offer a richer experience for him as a learner?

If you brought in a collection of things to class, what might you bring? Would your collection fit in your hands?

Teacher's note: How can we, as educators, move from a "count this" mentality to a "how many do you see?" mentality where there is more than one way to see math? Could incorporating this idea provide a more meaningful connection to counting for children rather than only using prescribed counting activities?

**BIG IDEA**   The Importance of Estimation

Which student's guess or estimation do you think might be the closest to the actual marbles Miss Penny is holding in her hands?

Were any estimations or guesses about the quantity of marbles too high? Were any estimations or guesses about the quantity of marbles too low?

Teacher's note: actively engage students in the book by asking them how many marbles they think are in Miss Penny's hands.

**BIG IDEA**   Counting by Ones

Use the strip of marbles pictured at the bottom of the page, below the illustration, to count the marbles as a class.

**BIG IDEA**   Conservation of Numbers/Quantitative Value of Numbers and the MP Standards in Action.

Will the number of marbles change, now that they are in the big space at the bottom of the basket? How many marbles do you think are at the bottom of the basket?

Teacher's note: On this page and throughout the book, Miss Penny says, "Prove it." This is a call to action that directly connects to mathematical practice standards. We want children to be able to communicate their understanding and persevere in problem solving. Asking students to "prove it" as a regular course of action, regardless of whether their answer is correct, is a great way to encourage this practice in the classroom.

---

**BIG IDEA**  Counting Strategies

Describe the different ways the students represented the number sixteen.

---

**BIG IDEA**  Same Value/Multiple Representations

Look at the representations of the number sixteen offered by the illustrator and those created by students on the final page of the book.

How can sixteen have the same value if it looks different in the pictures?

If you were to draw a representation of the number sixteen, what might it look like? Can you describe to the class what you would draw?

---

## IDEAS FOR AFTER THE STORY Retell, Recreate, and Talk about Math

A powerful way to use Miss Penny Says, "Prove It!" is to encourage children to retell Bobby's story using their own marble experience as a backdrop for understanding. Students reach into a marble jar and pull out their own collection of marbles to work with. Using the story board to guide them, students estimate, count one by one, explore ideas about the quantitative value of numbers, represent numbers in multiple ways, and hopefully gain an understanding of the conservation of numbers. These rich experiences allow children to navigate their own learning by exploring concrete, representational and abstract opportunities that promote understanding.

### Story Board Connection

Download the companion story board free at www.mathmilemarkers.com.

### Share Your Stories

Math MileMarkers® stories are the perfect way to enhance a child's understanding key concepts and have fun with math! Visit us at www.mathmilemarkers.com or on Twitter @mathmilemarkers to share how you used this Math MileMarkers® story to support learning in your class or at home. We love seeing pictures of students at work and hearing about the learning that happens when children retell, recreate, and talk about math with friends.

# ABOUT THE AUTHOR

**Lynda Brennan** is a 2020 recipient of the Presidential Award for Excellence in Mathematics and Science Teaching. She has over 25 years experience as a NYS classroom teacher and math specialist. In 2014, she launched Math MileMarkers®, a series of picture books designed to help children, parents and teachers enjoy, talk about and conquer the elementary math curriculum. New celebrated teachers/authors continue to join her mission adding to this collection of highly enjoyable, math-infused stories. For more information about the Math MileMarkers® series or learn about upcoming events, please visit www.mathmilemarkers.com.

# OTHER BOOKS AND PRODUCTS IN THE MATH MILEMARKER® SERIES

**Jayla the Number Navigator:** A Math-Infused Story about navigating the number line using combinations of 10. *Written by Lynda A. Brennan*

**On My Way to Grandma's House:** A Math-Infused Story about the number line and the concept of rounding. *Written by Lynda A. Brennan*

**Louie the Lucky Looker:** A Math-Infused Story about division. *Written by Marianne V. Strayton*

**Miss Penny Says Prove It!:** A Math-Infused Story about developing number knowledge and exercising the standards for mathematical practice. *Written by Lynda A. Brennan*

**Charlie in Fraction City:** A Math-Infused Story about understanding fractions as part of a whole. *Written by Lynda A. Brennan*

**Math MileMarkers® Foldable Stories:** A collection of Math-Infused mini-books for young learners. *Written by Lynda A. Brennan, Marianne V. Strayton, Scott Schaefer*

Please visit us at www.mathmilemarkers.com for curriculum companion activities for this and other Math MileMarkers® stories.

Made in United States
Orlando, FL
26 December 2024